PROJECTS IN PHYSICS

The books in the SCIENCE IN ACTION series are

Light and Sound

The Living World

The World of Numbers

Fun with Chemistry

Projects in Physics

Experiments in Physics

SCIENCE IN ACTION

The Marshall Cavendish Guide to Projects and Experiments

Compiled and revised by
Sue Lyon

PROJECTS IN PHYSICS

Projects created by
Paul Berman and Keith Wicks

Marshall Cavendish Corporation · New York · London · Toronto · Sydney

Editorial Staff

Series Editor	Sue Lyon	**Production Manager**	Carol Milligan
Assistant Editors	Nigel Rodgers	**Managing Editor**	Alan Ross
	Caroline Macy	**Editorial Director**	Maggi McCormick
Art Editor	Kay Carroll	**Publishing Coordinator**	Robert Paulley

This Edition published 1993

Published by Marshall Cavendish Corporation
2415 Jerusalem Avenue
North Bellmore
New York 11710

Typeset by Quadraset Ltd.
Printed in the USA by Worzalla Publishing Company, Wisconsin.

Science in Action derived from material previously published as Volume 25, Growing Up With Science, by H.S. Stuttman Inc.

© Marshall Cavendish Limited
MCMLXXXVII, MCMLXXXVIII

Library of Congress Cataloging-in-Publication Data available upon request
ISBN 0-86307-020-5(set)
ISBN 0-86307-941-5(vol)

Contents

Projects marked ⊕ need adult supervision.

Introduction

This book will help you learn more about science and technology. It includes experiments, projects, puzzles and even some tricks. Some of the experiments and projects are very easy. Others are a little harder (they are marked ⊕) and you will need help from one of your teachers or parents. You don't have to begin on page 8—look through the book and start with something you like—but remember that good scientists
- make a record of their work
- have a clean and tidy laboratory
- most important, keep themselves safe (always read pages 40 to 42 before you begin).

SIMPLE PROJECTS

Physics is the study of the natural phenomena of the universe, including the properties of energy, matter, light, gravity and magnetism.
In the following pages there are ten simple projects that use the laws of physics: make a boat race game, marbled paper or a model diver, learn about roller printing or magnetic fields, or see how you can make a piece of paper with only one side. (Remember that although these projects are easy to do, in some cases you may need help from an adult.)

Moebius bands

1 long thin piece of paper
glue
invisible tape
scissors

A regular sheet of paper has two sides. However, it is possible to make a sheet of paper with only one side. You will need a long thin piece of regular paper, such as a strip of paper from a cash register or a single column width cut from an old newspaper. You will also need some glue or invisible tape and a pair of scissors.

Procedure

1. Lay the **paper strip** flat on a table top.
2. Lift one end of the strip and bend it back, without creasing, until it touches the middle of the strip.
3. Lift the other end of the strip and bend it back until it touches the first end.
4. Twist one end of the strip one half turn so that the surface which was on top is now underneath (this is called a half twist).
5. **Tape** or **glue** the two ends of the strip together.

The resulting object may look as if it still has two sides and two edges but appearances can fool you. If you try to color just one side you will discover that you need to color the entire object. Similarly, you can show that the object has only one edge. One-sided, one-edged surfaces are called Moebius bands.

What do you think will happen if you cut a Moebius band in half along its length? Check your answer by cutting the band along its center line. Count the number of half twists in the resulting object. If the results surprise you, think hard before trying the next exercise. Make another Moebius band. You will be cutting this one along its entire length a third of the way from one edge. When you have worked out what you think will happen, make the cut with the scissors and see whether you were right.

Try making bands with more than one half twist in them. Check how many sides and edges each has, what kinds of object are formed when they are cut and whether these objects are linked together. After you have made and tested a few different bands, you should be able to discover some general rules about how Moebius bands behave.

Another project to try is making nested bands. Stack two or three identical strips of paper and make a Moebius band with one half twist from them, taking care to glue the correct pairs of ends together. Shake the resulting object to open it out. Check how many half twists each part of the object has.

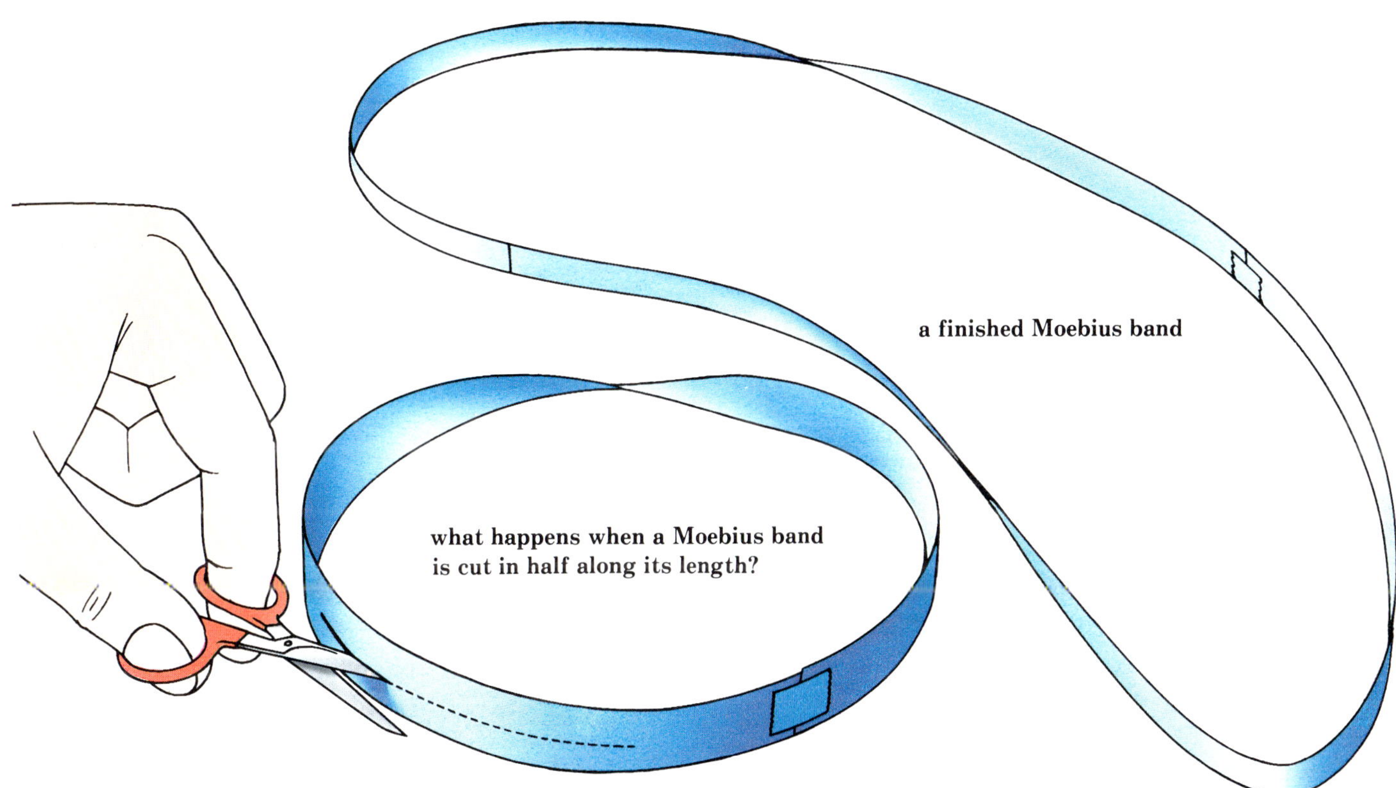

a finished Moebius band
what happens when a Moebius band
is cut in half along its length?

Marbling

You will need—

3 or more oil-based
  printing inks
old saucers
1 old plastic bowl
wooden popsicle sticks
strong brown paper
poster paper
turpentine
artist's paint brush
metal comb

The patterns that appear in marble can be recreated on paper by using inks that float on water. With a little care you can make designs that are very close to those of the real thing. Use the finished marbled paper to wrap up presents for your friends, or to make jackets for your books.

Procedure

1. For this job you will need at least three different-colored **oil-based printing inks**. Ordinary ink will not work. Also you need some **old saucers**, an **old plastic bowl**, and **wooden popsicle sticks**. It is very important that none of the saucers or the bowl will be used in the kitchen again.

2. Cover the work area with **strong brown paper**, or a thick layer of old newspaper as things will get quite messy.

3. Cut some **poster paper** into sheets that will fit inside the plastic bowl. Put the cut sheets to one side for the moment.

4. Pour some cold water into the bowl to a depth of about 4 inches (10 centimeters).

5. Squeeze out about a 1½ inch (3.5 centimeters) blob of ink onto a saucer and add about a teaspoonful of turpentine. Stir the mixture with the popsicle stick until both ink and turpentine are well-mixed. Repeat this procedure for the other colored inks.

6. Dip an **artist's paint brush** into one of the ink/turpentine mixes, then touch the surface of the water with the end of the brush. The ink should spread out from the end of the brush. If it does not, the mix is too thick —add a little more turpentine and mix it in well.

7. Wash out the brush, then repeat step 5 for the other two inks.

8. Now mix the colors into streaks by gently moving an old **metal comb** across the surface of the water. This will give a good "marbled" effect.

9. Gently lay one sheet of poster paper on the surface of the water and leave it there for a few seconds. The ink will be deposited on the paper.

10. Slowly lift off the paper, turn it right side up, and lay it on the brown paper to dry.

11. Try experimenting with different patterns, using the comb to sweep the colors into other designs. Once you have made a few designs on one side of a sheet, try doing a few double-sided ones. To get wavy patterns, pull the paper through the water rather than laying it on—wiggle the paper as you do it.

Take care!

Printing ink stains easily, so cover the work area (it might be best to work outside) and wash your hands at once if they are splashed with ink.

Static electricity

The electricity that drives the machines in your home is made by huge generators in power stations. You can generate small amounts of electricity without any special equipment. The electricity you make will be far too weak to make the smallest bulb light up, but you can use it to move small objects such as scraps of paper.

Generating static electricity

One easy way to generate static electricity is to rub a blown-up **balloon** on a piece of **cloth**. Wool cloth is ideal for this—try rubbing the balloon on a wool sweater or the sleeve of a wool jacket. Rubbing the balloon like this will charge it with static electricity. The charge should be strong enough to make the balloon cling to the wall for a minute or two.

You can show the power of static electricity in other ways. If you rub the body of a **plastic pen** on a piece of cloth, the pen will be charged with static electricity. You can measure the strength of the charge by tearing up a piece of **paper** into different-sized scraps. Rub the pen on different fabrics, such as wool, cotton and synthetic fibers. Now see which type of cloth gives the most charge by using the pen to lift up the scraps of paper—the bigger the scrap, the stronger the charge.

Conductors

Some substances, including water, serve as conductors of electricity—that is, an electric current can flow through them. You can prove this by running a slow stream of water from a faucet. Make the stream as slow as it will go before it begins to drip. Now, charge your plastic pen by rubbing it on some cloth and bring it slowly toward the stream of water, just below the faucet. The static electric charge will pull the stream of water toward the pen, but if you go too close and the pen touches the water, the electric charge will be conducted away by the water.

Boat race

You can use static electricity to play a simple "boat race" game. Each player needs a small, light **boat** and a **plastic pen**. If you have not got boats, use bottle **corks** instead. Float the boats on a large bowl of water. Charge your plastic pens and quickly push your boats under the water so that they come up wet. Use your pens to pull the boats from one side of the bowl to the other. If your pen touches the wet boat or the surface of the water, the charge will be conducted away and your boat will stop.

Did you know . . .

Benjamin Franklin, one of the signatories of the Declaration of Independence and the first American ambassador to France, was first and foremost a scientist. His most famous experiment concerned static electricity. In 1746 he saw a demonstration of static electricity at Edinburgh, Scotland, and bought the equipment to conduct his own experiments. He figured out the idea of static electricity by watching how it jumped from one of his colleagues to another in 1747. He also invented the lightning rod in 1752.

FINISH
if your pen
touches the boat,
the charge will be
conducted away
use a charged pen
to pull the boat

Magnetic fields

1 metal file
1 old lump of iron
1 vise
1 horseshoe magnet
1 knitting needle
iron filings
cardboard
container with plastic lid
aluminum foil

The poles of a magnet have the strongest magnetic effect. While the middle of a bar magnet has little or no power, the ends can easily attract small pieces of iron or steel. You can demonstrate this by making a magnet and some iron filings.

Procedure

1. To make iron filings, all you need is a metal **file** and an old **lump of iron** such as a nut, bolt or nail. Be careful not to file your fingers when you are filing the iron. If you have a work bench with a **vise**, it is probably best to use it to hold the metal while you work on it.

2. When you have enough filings, prepare the magnet. Any iron or steel object can be made into a magnet. All you need to do is run an ordinary horseshoe **magnet** along the chosen object and it becomes magnetic. If the object is made from iron it will not stay magnetic but a steel object such as a knitting needle will be permanently magnetized.

3. If you now pass the steel magnet over the iron filings, you will see many filings clinging to both ends of the magnet and hardly any to the middle. There is, however, an area around a magnet where the influence of the magnet can be felt—this is called a magnetic field. It is invisible but you can show that it exists in this simple experiment. You will need a bar magnet, a piece of thin cardboard, and iron filings (you will need quite a lot of filings to do this properly).

Magnetic fields

Place the cardboard on top of the magnet—have the magnet under the center of the cardboard. Then, using a paper scoop, sprinkle filings onto the cardboard. When you tap the sides of the cardboard, the filings will arrange themselves along a series of lines. These lines will always be the same no matter how many times you try this. Try also arranging four bar magnets under the cardboard just far away enough from each other not to attract or repel, and sprinkle the filings onto the cardboard. Tap the cardboard and see the patterns that are made.

Permanent patterns

It is possible to make a permanent record of the patterns made by magnetic fields and iron filings. If you wish, you can color the iron filings. You will need a container with a plastic lid, the filings and a quick-drying spray paint. Cut a hole in the lid just large enough to spray through. Place the filings in the container, put the lid on and give a quick spray. Transfer the filings to some aluminum foil to dry, but do not let them stick. To make the picture, take another sheet of cardboard and some transparent glue. Spread the glue on the cardboard. Make the magnetic field pattern in the same way as before, then put the glued cardboard, glue side down, onto the filings. Lift the glued cardboard away, and the filings should stick to it.

Take care!

Be very careful that, when filing the old iron, you do not hurt your fingers. If you have one, place the iron in a vise–you will find the job much easier.

Newton's law

You will need—

2 rulers or 2 long straight pieces of wood
selection of coins, at least 6 having the same value
double-sided tape or modeling clay

You have probably seen a Newton's cradle, where a number of steel balls are suspended from a frame. When you pull one of the balls back and let it go, it hits the row of balls. But, rather than the whole series of balls moving, only the one on the end shoots off.

You can do this type of trick yourself using two long rulers or straight pieces of wood, and a selection of coins. Make sure that at least six of the coins are of the same value—the rest can be anything you like.

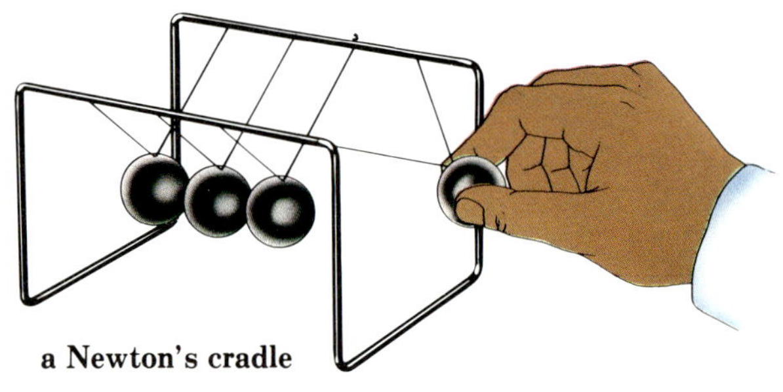

a Newton's cradle

Procedure

1. Place the two **rulers** on a surface that the **coins** will slide on easily. A laminated plastic counter or table is ideal for this. Make sure the rulers are parallel to each other, and use **double-sided tape** or a piece of **modeling clay** to keep each one firmly in position.
2. Lay five identical coins in line between the two rulers, making sure that each one touches the next. Arrange them so that they are about one inch (25 millimeters) in from one end of the rulers (see illustration).
3. Now place another identical coin at the entrance to the gap between the rulers.
4. Carefully flick that coin toward the coins between the rulers, making sure it hits them dead center. The whole group will move a little, but the end result will be that the coin on the other end flies off.
5. Repeat the experiment several times, flicking the coin harder or softer and watching the results. The coin on the end will fly off more or less depending on how hard the first coin is flicked.
6. This phenomenon is explained by Newton's third law which says that to every action there must be a reaction. When you flick the coin, it hits the first one (the action) and that coin then tries to move away from the first one (the reaction). But it can't move because it is prevented from doing so by the next coin in the line. So, the force of the impact is passed on to the next coin until it gets to the end of the line. At this point there is nothing preventing the last coin from moving, so it flies off.
7. In theory, the row of coins shouldn't move when hit. But in practice this isn't so, as you will notice.
8. But the coins do not have to move in order to pass on the force. Ask a friend to hold down one of the coins in the row (not the one on the end) and try flicking the coins again. You'll find that the one of the end flies off as before—the force is still passed on.

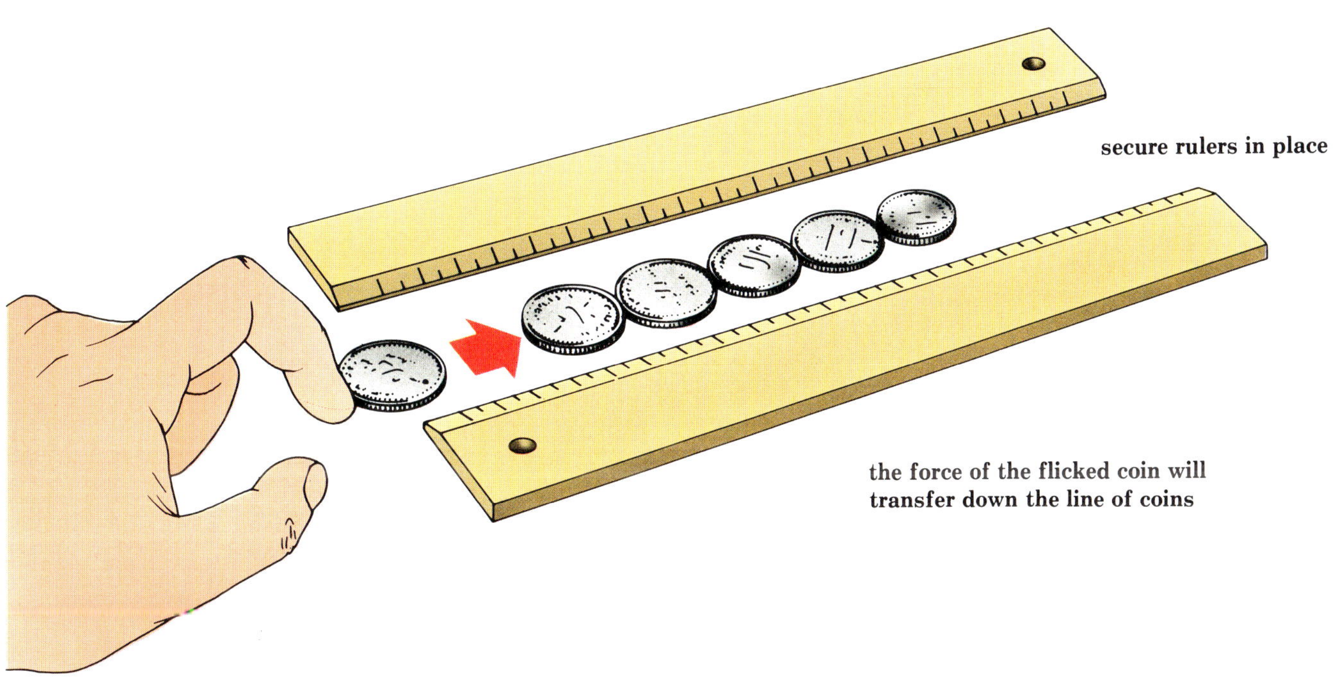

9. Now try flicking two identical coins against the row to see what happens. This time two coins will fly off the end. No matter how many identical coins you knock into the row, the number of coins flying off will be exactly the same.

10. Also, you can do the experiment with coins of different weights. If you flick a heavy coin against a row of identical coins which has a light coin at the end, you will find the one at the end flies off very quickly, followed by one of the heavier coins.

Plumb line

You will need–

thin string
weight (1 small stone or
  pebble)
1 paper clip
modeling clay
1 thumb tack

Take care!

When testing your plumb line, use a step ladder to attach the line to the door frame, or ask an adult to do it for you. Never use a stool or chair–you could fall.

A plumb line is simply a line with a weight attached to one end, and by using it you can check whether things are vertical. Although you can buy a plumb line from a hardware store, it is very easy to make one. All you need is about 6 feet (180 centimeters) of **thin string**, a **weight** such as a small stone or pebble, a **paper clip** and a little **modeling clay**.

Procedure
1. Tie the paper clip to one end of the string. Hold the weight on the clip and mold the modeling clay around both of them.
2. You can test your plumb line on a door frame (but make sure the door is open). Using a **thumb tack**, fix the line to the top of the frame and wait until the weight stops swinging. If there is a large gap between the frame and the bottom of the plumb line, the frame is not vertical; it is leaning inward.
3. If the weight hangs against or near to the frame, the frame is not vertical; it is leaning outward.

A useful tool
The plumb line shows the true vertical because of the force of gravity. When dropped, all objects fall vertically. The weight cannot fall to the ground because it is held by the line, but as it hangs, it pulls on the line so that it is vertical.

  Plumb lines are useful for any job where you need to know the true vertical. For example, use one to hang your pictures to make sure that they are not crooked.

Spirit level

Just as a plumb line is used to check the true vertical, so a spirit level is used to make sure that things are exactly horizontal (level with the horizon). To make this simple spirit level, you need a **glass jar with a screw top**, some **colored liquid** (you could dye water with food coloring, or use cold coffee or tea) and some **colored adhesive tape**.

Procedure

1. Pour the colored liquid into the jar so that it comes just below half way. Screw on the top of the jar. Put the jar on the edge of a flat surface like a table. Hang your **plumb line** next to the jar, but without touching it. The side of the jar should be parallel with the plumb line; if necessary, wedge the jar with scraps of **paper** or pieces of **wood**. Fix the colored adhesive tape to the jar so that it is exactly on the line made by the surface of the liquid.

2. If the jar is placed on a surface that is not horizontal, the liquid will not line up with the edge of the adhesive tape.

Water level

This spirit level depends on the fact that liquids like water always find their own horizontal level. Try gently tilting the spirit level and you will see that the water moves so that the surface is always horizontal.

Sloping shelves will not hold your possessions safely, so use your spirit level to check that your shelves are horizontal.

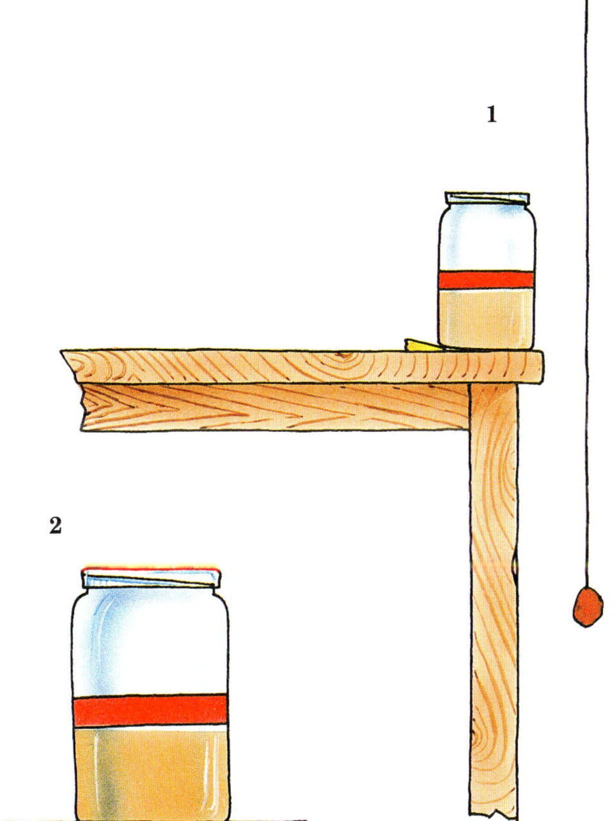

You will need–

1 glass jar with a screw top
colored liquid
colored adhesive tape
plumb line
small amounts of paper or
   wood

Model diver

You will need—

1 clear glass wine bottle
1 tight-fitting cork
1 piece of flat, rigid plastic
1 plastic tube from an old
  ballpoint pen
waterproof glue
thin wire
scissors
modeling clay
oil paints

At your command, this model diver will move up or down in the water. You can even make the diver hover at any depth you choose.

Procedure
1. Obtain an empty **wine bottle** made of clear glass and with a tight-fitting **cork**. You will also need a flat piece of thin but rigid **plastic**, the **plastic tube** from an old ballpoint pen refill, some **waterproof glue**, a few inches of **thin wire**, and a pair of **scissors**.
2. If the cork is dry and rigid, leave it in water until it becomes flexible.
3. Draw the outline of the diver on the flat plastic sheet. Be sure to make it thin enough to fit into the wine bottle. Use scissors to cut the diver to shape. If you wish, you can add detail to the diver using **oil paints**.
4. Take a small piece of the plastic tube and seal one end and the breather holes with modeling clay. The other end should remain open. Glue this tube to the diver, as shown in the diagram. The air in the tube will make the diver float.
5. Wind some wire around the diver's feet. The weight of this wire will make the diver stay upright in the water. Use enough wire to make the diver sink. Then

remove a few turns at a time until the diver just floats. It is now ready to use.
6. Put the diver in the wine bottle and almost fill the bottle with water. The diver should float.
7. Push the cork a little way into the bottle. Doing this will increase the pressure in the water and force a little of it into the tube on the diver. As a result, the diver sinks.
8. Pull the cork out a little to reverse the process and make the diver rise. With careful adjustment of the cork, the diver can be made to stop at any depth.

Did you know . . .

In real life, divers have to be extremely careful about the speed at which they come back to the surface to avoid what is called "the bends." This is a crippling sickness that can kill deep-sea divers. As divers descend, the pressure builds up in the air they breathe. This does not matter until they start their climb back, when bubbles of air can form in their blood, due to the lowering of pressure. Too much air in the blood is fatal. To avoid this, divers rest at different levels in "decompression chambers" until they are ready to go up further.

controlling the diver

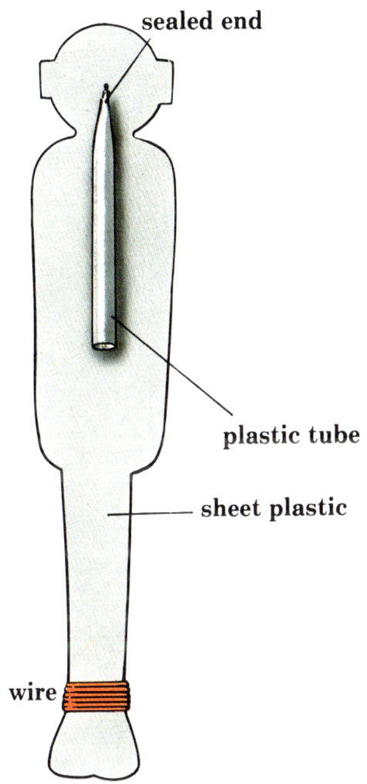

back view

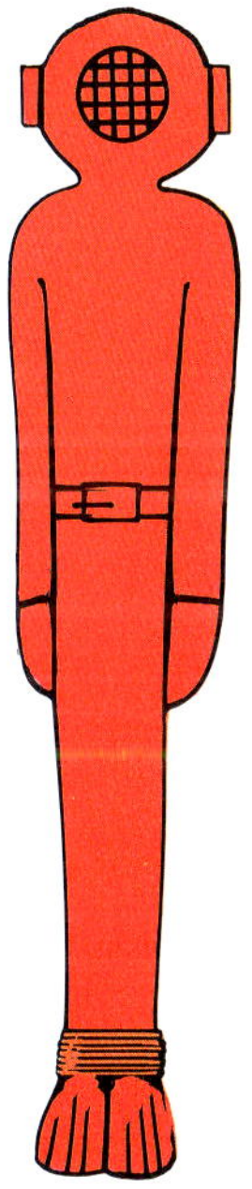

front view

Take care!

Use a hacksaw to cut the plastic tube to size. It will be easier and safer if you use a vise to hold the tube firmly.

Roller printing

You will need—

1 measuring tape
lining paper
1 tin can
felt
1 knitting needle
glue
1 pair of scissors
1 long nail
1 metal or plastic tray
colored paint
1 hammer

Roller printing is an easy way of getting a repeated pattern onto paper. There are several uses for this type of printing. For instance, many wallpapers are produced using this method. A simple roller can easily be made at home and used to produce a variety of decorative pictures. One good use for roller printing at home would be to make a border or frieze for your own bedroom (but make sure you have permission before you stick it up).

The equipment you will need for this project includes a **measuring tape**, suitable **paper**—lining paper would do, or you could use the back of some unwanted wall paper (but not the textured, embossed or prepasted types), a **tin can** or stout, cylindrical cardboard container with a tight-fitting, removable lid, thick **felt** or other thick material, an old **knitting needle** long enough to reach through the can and still leave enough sticking out at both ends for you to hold in your hands, suitable **glue**—check the glue label to make sure it will stick fabric to metal—**scissors**, a **hammer**, a long **nail** of slightly narrower diameter than the knitting needle, a shallow metal or plastic **tray**, and colored **paint** of your choice. You will also need a large flat surface to work on, such as a work bench or even the floor in a garage or shed—make sure the surface is clean and cover it with newspapers to protect it from the paint.

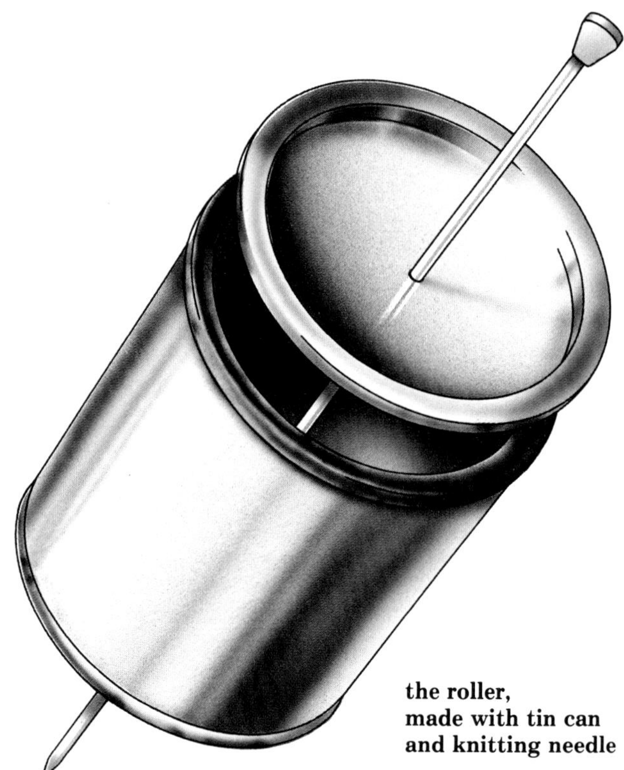

the roller,
made with tin can
and knitting needle

Making a Border

1. The first step is to measure how long the border needs to be and decide how deep it should be. Then, measuring carefully, cut out the right size piece of paper. If you need a long border, you may find it easier to cut the paper into more manageable strips of 3 feet to 6 feet (1—2 meters) long.
2. To make the roller, mark the centers of the bottom and lid of the can and, using the nail as a punch, tap it lightly with the hammer to make a hole at each end.
3. Thread the knitting needle through the hole in the bottom and then through the lid—making sure the lid is on tight. The knitting needle should be a tight fit in the holes. For safety, put some tape over the point of the needle. If you do not have suitable equipment to make your own roller, an old rolling pin can be used.
4. Now cut the felt into the shapes you want for your border and glue them onto the can. Make sure the shapes are lined up properly in relation to the ends of the can and to each other. If you are using a repeat design of different shapes, mark the shape that comes first in the sequence.
5. Next, you should put a thin layer of paint in the bottom of the shallow tray and roll the shapes in it. If you do not have a suitable tray, keep the paint in a small can and cover the shapes using a paint brush.

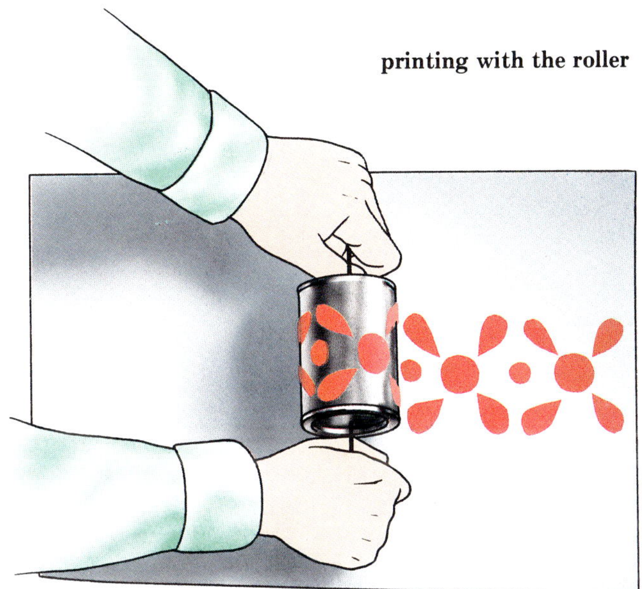

printing with the roller

6. After loading the shapes with paint, run the roller along the strip of paper—holding the paper down firmly while you do it and refilling with paint when necessary. When the strip is full, leave it to dry. It will then be ready to put up on the wall.

Take care!

When making the holes in the can, hold the nail well away from its top to avoid hurting your fingers with the hammer.

Parachute

You will need—

1 square silk or nylon scarf
2 pieces of string
1 small weight or toy
  soldier

If you were to drop a feather and a lead weight, you would be surprised if they reached the ground together. Yet this is what would happen if you could perform the experiment in a vacuum. For it is only the resistance of the air to movement through it that causes some objects to float gently to the ground. Try making this model parachute and see how it uses air resistance to ensure a slow descent.

Procedure

1. Use a **square silk or nylon scarf** for the canopy of the parachute. A large, thin scarf will work best.
2. Cut a piece of **string** so that it is about three times the length of one side of the scarf. Tie the ends of the string to two adjacent corners of the scarf. Then take another piece of string the same length and tie it to the other two corners.
3. Tie the two strings to a **small weight**, as shown in the diagram. You could try using a toy soldier as the weight. The precise weight required will have to be found by trial and error, as it depends on the size and weight of the parachute.
4. Loosely fold the scarf and place the weight on it. Then throw them high into the air. The weight should pull the parachute open as it falls, and so descend slowly to the ground. If the strings become badly tangled, untying them from the weight will make them much easier to sort out.

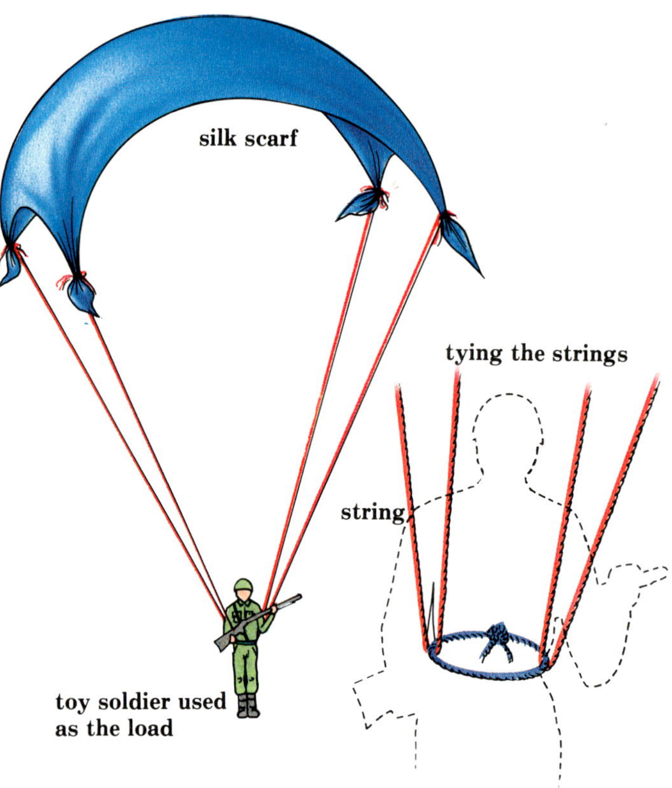

ADVANCED PROJECTS

In the following pages, there are eight more projects that use the laws of physics. These projects are fairly easy, but you may need to buy more unusual equipment and, for some projects, it would be safer to have either help or supervision from an adult. However, all the projects are fun to do: you can make your own sundial, rocket boat or a radio that works without electricity; you can test waves with your own machine or make pendulum patterns; and you can even push an egg into a bottle.

Egg in a bottle

1 hard-boiled egg
2 or 3 matches
1 clear glass bottle

When air is heated it expands, and when it cools it contracts again. This is the principle on which the hot-air balloon works. Heated air is fed in at the bottom of the balloon and it expands, filling the balloon. As the hot air is also lighter than the cold air around it, it creates lift to make the balloon fly.

If a balloon were to fly into bright sunshine, the air inside the balloon would heat up further and expand more, and there might be a danger of the balloon bursting. So a hot-air balloon has built-in safety valves that allow air to be released if the air expands too much.

If a balloon flies into clouds, the air inside cools and contracts, so more hot air has to be fed in from the gas burner at the bottom to keep the balloon in the air.

You can demonstrate that air expands and contracts using a hard-boiled egg and a wide-necked bottle.

Procedure

1. Choose an **egg** that is just too large to fit through the neck of the **bottle**. Remember to take into account the thickness of the eggshell, which will be removed when the egg is cooked.
2. Boil the egg for about ten minutes to make sure that it is completely hard, then place it in cold water to cool. If the shell splits and the white is deformed, the experiment will not work properly, so you will have to boil another egg. The egg must be in its smooth oval shape.
3. If you are not sure whether the egg is hard-boiled, place it on a hard table or counter and spin it. If the egg is still soft inside, it will spin cleanly. A hard-boiled egg will not spin neatly—it will wobble.
4. Peel the egg—take care not to break the egg white.
5. Now try the egg for fit in the neck of the bottle. It should be just too large to slip through. Remove the egg from the neck.
6. Light two or three **matches** and drop them, still burning, into the bottle.
7. Let them burn out, then quickly place the egg, thinnest end pointing down, into the neck of the bottle. Keep a finger resting lightly on the egg to insure a good seal with the bottle.
8. As the air inside the bottle cools it will contract, pulling the egg into the bottle.
9. If you do not want to use matches to heat up the air, another method you can try is to place the base of the bottle in hot (not boiling) water. Once the air is warmed, place the egg over the bottle's neck and, again, it will be drawn in.

Did you know . . .

The very first flight ever made by human beings used the principle of hot air rising to give it lift. On November 21, 1783, the paper balloon built by the Montgolfier brothers sailed into the sky from Versailles, France, and carried two men—Pilâtre de Rozier, a scientist, and the Marquis d'Arlandes, an army officer—5½ miles (9 kilometers) in 25 minutes. In 1785 John Jeffries, an American doctor, and Jean-Pierre Blanchard, a Frenchman, crossed the English Channel in 2 hours. In 1793, Blanchard made the first flight in North America, observed by President George Washington.

Take care!

After taking the pan off the heat, remove the egg from the boiling water with a large slotted spoon. If you are splashed with the very hot water, rinse your hand in cold, running water until the pain stops. Then tell an adult at once.

Pendulum patterns

You will need—

1 weight
1 piece of string
1 funnel
dry sand or paint
sheet of paper
adhesive tape
liquid glue
powder paint, cereal
    grains or small sea shells

You can make a simple pendulum by hanging a **weight** on a piece of **string** from an open doorway, a tree branch or a stepladder. By pushing the weight in different directions, you can make it swing in a straight line, an ellipse or a circle. Try experimenting with different weights and lengths of string. You should find that a long pendulum swings more slowly than a short one, and that varying the weight has little effect on the swing. Make a really long pendulum so that you can see more easily exactly how it swings. Also, you can place a sheet of paper underneath it on a sunny day and trace the path taken by its shadow. A better way of looking at the movement of a pendulum is to make it draw its own path for you. You can do this by using a **funnel** with a hole in its base as the weight. The funnel can be filled with dry **sand** which will leave a trail as the pendulum swings. You may need to try a few different hole sizes in order to get a steady stream of sand from the funnel. The funnel should be completely dry, or the sand will stick to it. You can use tape to block off part of a hole that is too large. With a little practice, you should be able to produce smooth lines, ellipses and circles of different sizes on a sheet of paper placed under the pendulum. Make sure that you start the pendulum moving smoothly, otherwise you will get an uneven pattern. You will also need to uncover the hole at the right time, or you may get a large pile of sand before the

a pattern made with a two-section pendulum

Did you know . . .

The invention of the pendulum in the 17th century made possible truly accurate clocks for the first time. Christiaan Huygens, the Dutch physicist, figured out the right length of pendulum and how to hang it to control a clock's mechanism.

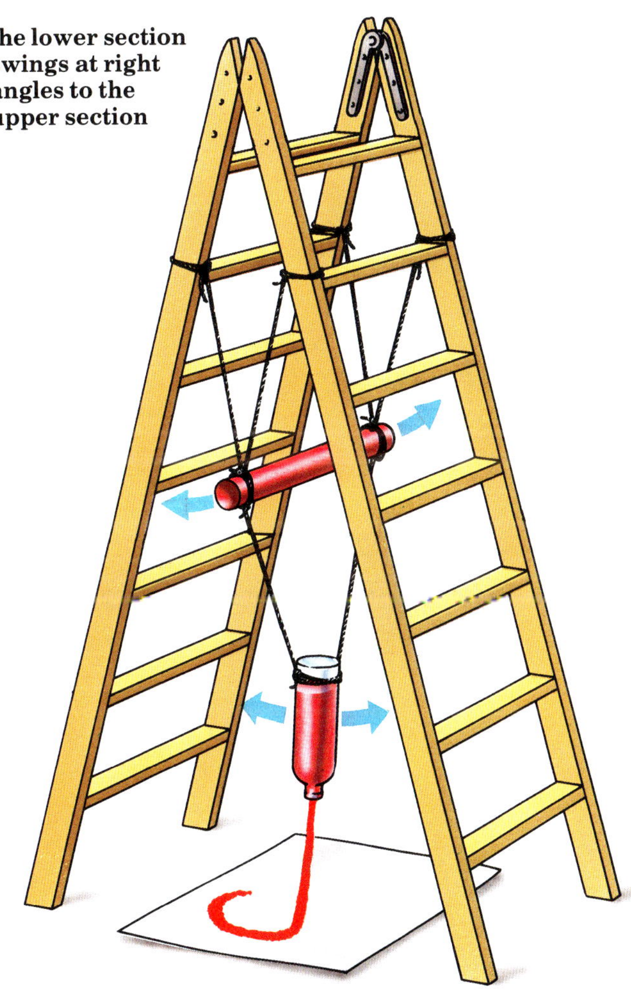

pendulum starts to move. You could use liquid paint instead of sand, but you would probably need to use a smaller hole and work in a place where no-one will object to paint being spilled. Another possibility is to use liquid glue in the pendulum and, after the pattern has been drawn, cover the paper with powder paint, sand, cereal grains or small sea shells which will stick to form a pattern along the path taken by the pendulum.

You will notice that the size of the pendulum swings gradually decreases. This is due mainly to air resistance and the string's resistance to being bent. If you leave the pendulum swinging in a circular path, you will see that it is actually following a long spiral rather than a perfect circle. You can reduce the effects of air resistance by making the weight heavier. You can increase air resistance by attaching vertical cardboard fins to the sides of the weight. See what effects each of these has on the patterns produced.

When you have mastered starting, stopping and controlling the flow rate of the simple pendulum, you are ready to use the more complex pendulum shown in the illustration. This pendulum consists of two sections; the upper section swings in one direction while the lower section swings at right angles to it. If the two parts of the pendulum are of different heights, they will swing at different rates producing a variety of complex patterns.

Take care!

When hanging a long pendulum, use a firm step ladder or ask an adult to attach the pendulum.

Wave machine

You will need—

1 small electric motor for
   12-volt supply
1 battery of lower voltage
1 strip of wood
rubber bands
1 empty thread spool
small pieces of wood
1 short wooden rod
string
1 cork

This simple machine enables you to study wave motion. See how waves travel through water, and simulate the build-up of shock waves on a fast-moving airplane. You will also be able to demonstrate why the sound from a moving vehicle changes pitch as it passes you.

Construction

1. You will need a **small electric motor** designed to run from a supply of about 12 volts. As the motor will run slowly, it will be connected to a **battery** of lower voltage. A 4½-volt battery should be satisfactory, but you may have to experiment to find the right voltage.
2. Fix the motor to the end of a **strip of wood**. Position the motor so that its spindle hangs over the end of the wood. If the motor cannot be screwed down, fix it to the wood with **rubber bands**.
3. Push the motor spindle into the hole in an empty **thread spool**. Jam the spindle in place with small **pieces of wood** so that it is off center.
4. Screw a **hook** or eye into the end of a short **wooden rod**. You can use a piece of bamboo if you first pack one end with scraps of wood.
5. Hang the rod on a loop of **string** passing over the spool and through the hook or eye.
6. Connect the electric motor to the battery. The motor should turn fairly slowly, making the spool wobble and the rod bob up and down.

Making waves

1. Run some cold water into a **bathtub** until it is about half full. Hold the wave machine over the tub so that the rod dips into the water. The up-and-down movement of the rod will cause a series of circular waves to spread out around it. Although these disturbances move outward, the water itself moves up and down. You can check this by floating a small piece of **cork** in the water a short distance from the rod. The cork will bob up and down without moving away from the rod.
2. Now make the rod move through the water as it bobs up and down. Circles of waves will still be formed. But each circle is centered on a different point. As a result, the waves are compressed (squashed together) in front of the rod, and spread out behind it.

This wave pattern shows why the Doppler effect occurs. The waves pass any point in front of the rod at a greater rate, or frequency, then they pass a point behind the rod. In the case of sound waves, this increase in frequency corresponds to an increase in pitch. This is why a train whistle seems to have a higher pitch when the train is approaching.

A similar wave pattern also forms in front of fast-moving aircraft. At high speed, the waves combine to form a shock wave. Try simulating this effect using the wave machine.

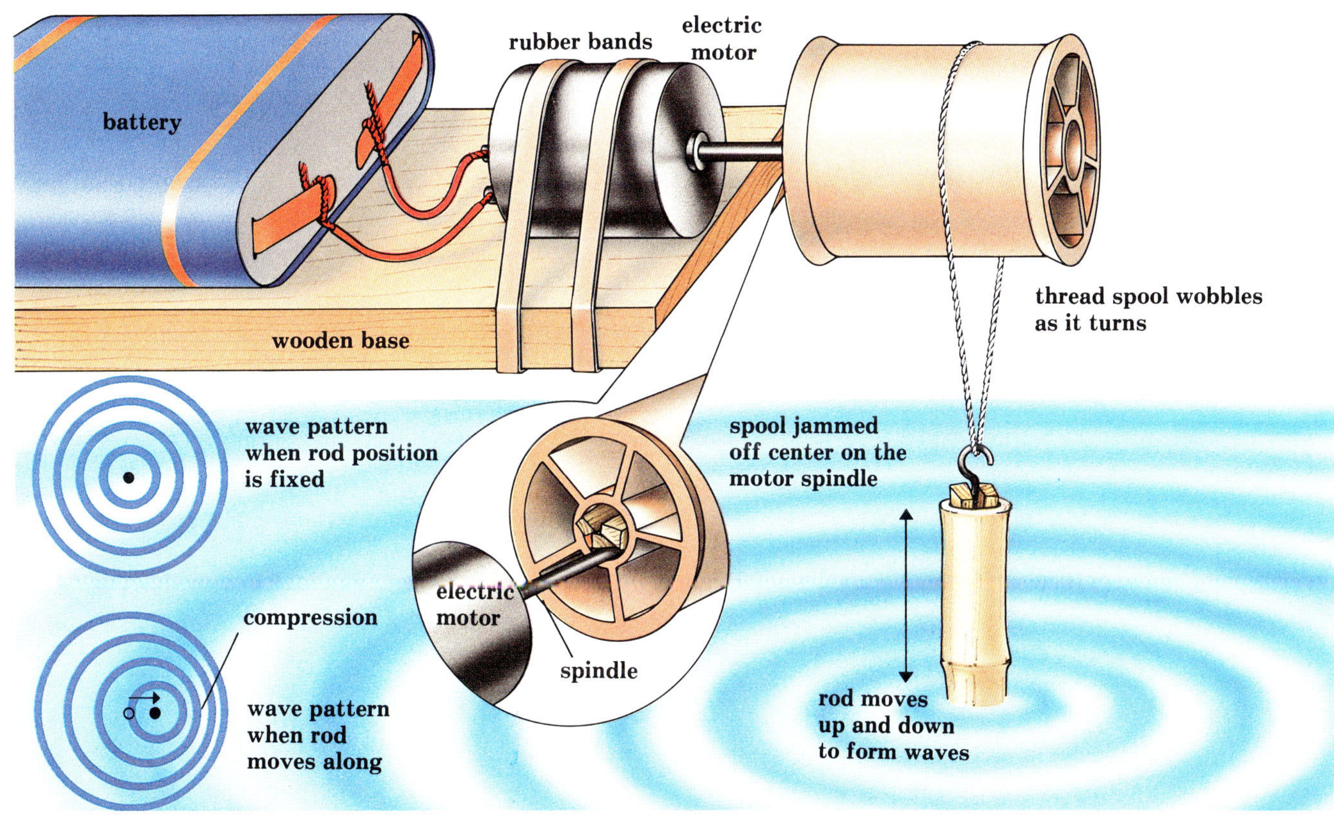

rubber bands
electric motor
battery
thread spool wobbles
as it turns
wooden base
wave pattern
when rod position
is fixed
spool jammed
off center on the
motor spindle
compression
electric
motor
spindle
wave pattern
when rod
moves along
rod moves
up and down
to form waves

Galvanometer

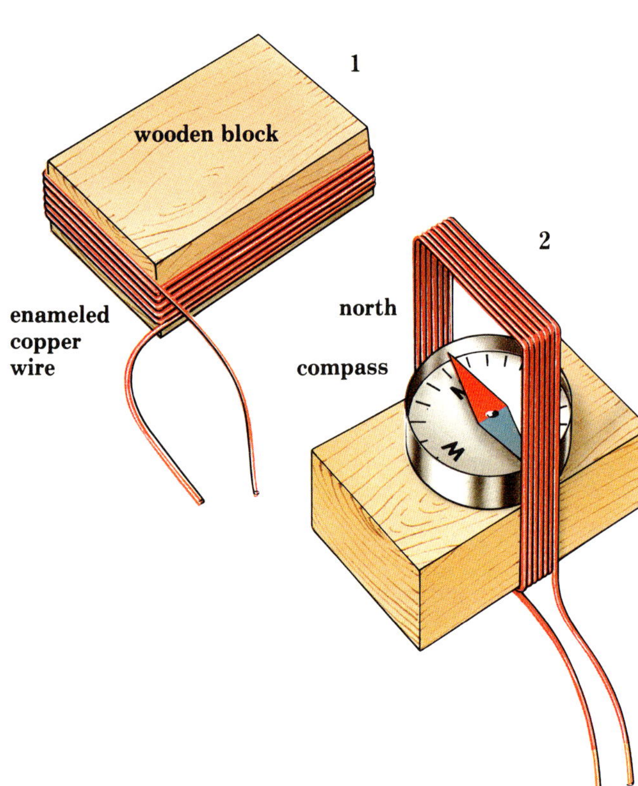

Use a pocket compass and a few yards of wire to make a galvanometer. This instrument uses the principle of electromagnetism to detect electric currents. Even a weak current flowing through the instrument will cause the compass needle to move.

Materials

Buy a small magnetic **compass** from a toy store, and a few yards of **enameled copper wire** from a radio parts supplier. This wire is available in various gauges (thicknesses). Gauge 18 wire is convenient, as it can be wound into a neat coil easily without much danger of breaking. However, almost any type of wire can be used, but it must be covered with enamel or some other form of insulation. The length of wire is not too important, but the galvanometer will not be as sensitive if too short a piece is used.

The only other materials needed to make the galvanometer are a small **block of wood**, **thread**, **fine sandpaper** and some **glue**. To demonstrate the action of the instrument, you can use any flashlight **battery** and a **miniature bulb** of the same voltage. Have a **holder** for the bulb so that wires can be easily connected to it.

Construction

To construct the galvanometer, proceed as follows:
1. **Saw** a piece off a strip of wood to make the base of

the instrument. The width of the strip should be the same as the distance across the compass. The other dimensions of the base should be roughly in the proportions shown.

2. Wind the wire loosely around the block of wood to form a rectangular coil (figure 1). Tie the windings together at various points with pieces of thread.

3. Carefully remove the coil, reposition it as shown in figure 2, and glue it to the block. Then glue the compass to the block, so that the North point on the compass card is positioned as shown.

4. Using the sandpaper, carefully scrape the enamel insulation from the ends of the coil. The galvanometer is now complete.

Testing the galvanometer

Turn the instrument until the compass needle lies directly under the coil (figure 3). Then connect the coil to the battery and bulb (figure 4). The coil will now act as an electromagnet and make the compass needle swing to the side. Reversing the battery connections (figure 5) will make the needle swing the other way. The bulb is used as a resistor to reduce the current flow and thus make the battery last longer.

You can use the galvanometer as a test instrument to confirm the flow of direct (one-way) current in circuits that appear in other projects.

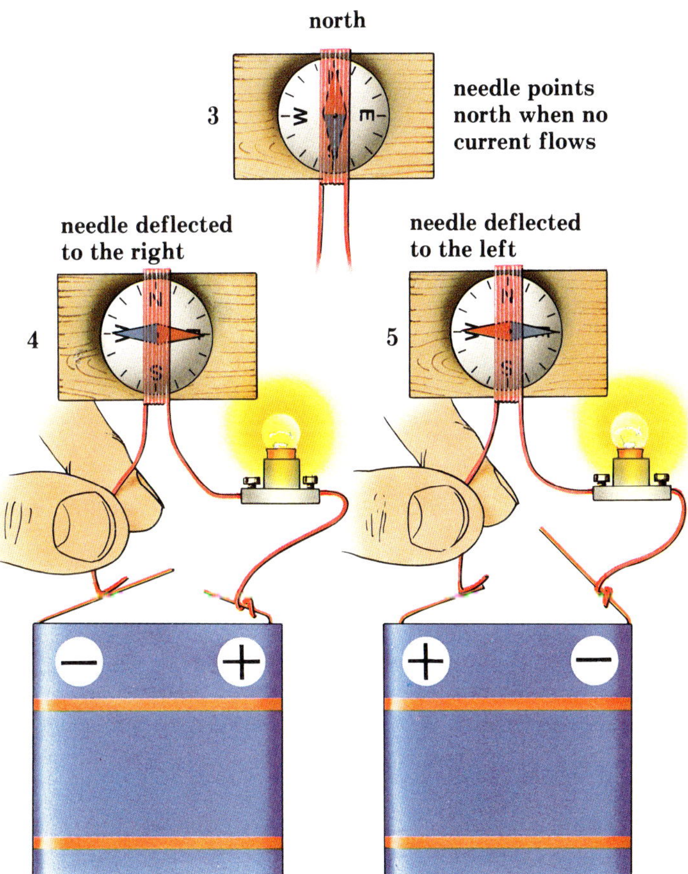

Sundial

Take care!

Try to find a jar with a cardboard lid which will be easy to puncture with the knitting needle. If your jar has a metal lid, make a hole by tapping in a nail with a hammer, but be careful not to hurt your fingers. Never try to make a hole in a glass lid.

Before the invention of mechanical clocks, people used sundials as one method of telling the time, and as long as the sun casts a shadow, a sundial is a fairly efficient time piece. See how efficient it is by making your own. You will need a large round **glass jar with a lid**, a 1-inch (2.5-centimeter) wide **strip of thin cardboard** long enough to fit snugly inside the jar, a **ruler** and **pencil**, some **glue**, a **knitting needle** about 2 to 3 inches (5 to 7 centimeters) longer than the jar, **modeling clay**, and some **wood** to make a stand or a wedge.

Procedure

1. Using a ruler and pencil, mark the cardboard strip in 24 equal spaces, then divide each space into four. Glue the strip inside the jar half way up and with the numerals facing inward.

2. Push the knitting needle through the lid and fix the point with modeling clay on the base of the jar.

3. Set the jar on a wedge of wood or make a stand like the one shown in the illustration. The tilt of the stand should equal 90° minus your latitude (check in an atlas).

4. Set the jar and its stand to face the sun when it is noon by a clock (or one o'clock during daylight saving time). Turn the jar until the needle's shadow falls exactly on the 12 mark. Now glue the jar to the stand.

5. The sundial now shows "sun time," so add one hour when daylight saving is in force. The needle will cast a shadow to indicate the number of the hour and any quarters (the four divisions of each numbered space).

Note that the sundial is a "twenty-four hour clock," so after noon, you will need to subtract 12 from the hour to figure out the time.

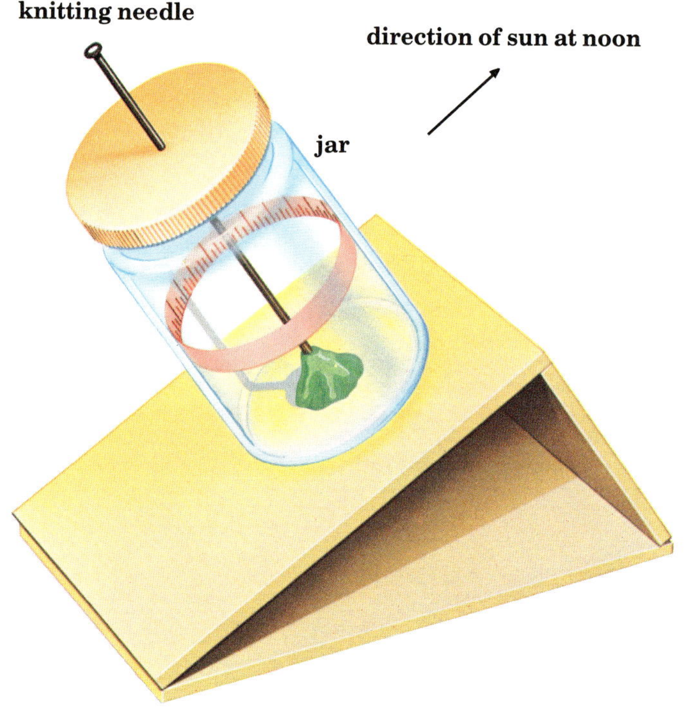

Rocket boat

The world's mightiest rockets like NASA's giant Saturn 5 work on exactly the same principle as an air-filled rubber balloon which flies around the room when you release it. They are both pushed forward by the backward rush of escaping gas.

In a rocket, the fuel is ignited and, as it burns, generates high-pressure gas, which pushes out in all directions inside the rocket (figure 1). The rocket does not move because the pressure is equally strong in all directions. However, when the rocket's exhaust is opened (figure 2), hot gas rushes out, creating an upward thrust, and the rocket is pushed from its launch pad. You can see how exhaust gas powers a rocket by making a rocket boat.

Procedure

1. Obtain a hollow aluminum tube with a screw lid; a **cigar tube** is ideal. Using a **nail** and **hammer**, punch a hole in the screw top. Twist a **pipe cleaner** around each end of the tube as shown in the illustration and **glue** into position.

2. Half-fill the tube with **water** and replace the lid. Stand the tube and three **candle stubs** in an **aluminum food tray** and float the boat in a bath.

3. Now (and **not** before) light the candles. After a few minutes, the boat will be thrust forward by a jet of exhaust steam (figure 3).

Take care!

After the candle stubs have been lit, never touch the hot metal tube or place your hand in the jet of steam. If you do accidentally burn your hand, cool it under running cold water and tell an adult immediately.

When making the hole in the tube top, make sure you do not hit your fingers with the hammer.

Propeller power

You will need—

1 small electric motor
1 battery
1 soft-drink can
strong scissors
gloves
1 block of wood
1 wood screw
1 rigid plastic tube
1 rectangular plastic
 container
1 small cardboard carton
1 rubber band
1 plastic propeller

Most boats are thrust forward by means of one or more propellers that rotate in the water. But the model boat described here uses the same form of propulsion as a hovercraft. It has a propeller that rotates in the air.

Construction

1. You will need a **very small electric motor** and a suitable **battery** for it. The motor must not be very heavy as it will be mounted high on the boat and could cause it to capsize.
2. Cut the propeller shape shown in the illustration from a piece of thin steel. You can get this from the side of a **soft-drink can**—they are usually quite thin and easy to cut with strong **scissors**. **File** down any sharp edges. Then twist the two blades of the propeller in opposite directions, as shown. Wear gloves to protect hands.
3. Make a hole in the center of the propeller by placing it on a **block of wood** and hammering a small round **nail** through it.
4. Now take a small piece of fairly rigid **plastic tube** to fit tightly over the spindle of the motor. You should be able to get a suitable piece of tube from the refill of an old ballpoint pen. Screw the propeller to one end of the tube with a **wood screw**. Then push the other end of the tube over the spindle of the motor.
5. A **rectangular plastic container** can be used for the hull of the boat. But use the hull of a toy boat if you have a suitable one. If you use a plastic box, bend a piece of **thin steel** to form a pointed bow section, as shown. Glue this to one end of the box. You will also need a simple rudder. A piece of thin steel bent at right angles will do.

Glue this to the other end of the box.
6. Place a **small cardboard carton** between the electric motor and the battery. Put a strong **rubber band** around the three items to hold them together. Then place them in the back of the boat. The carton is used to raise the motor so that the propeller is kept just clear of the water. Do not mount the motor higher than necessary as this will tend to make the boat top-heavy.
7. Run some water into a **bathtub** and float the boat in it. The weight at the back will make the boat tip up at the front. But this does not matter as long as the back is not pushed under the surface. If the boat tends to tilt sideways, shift the battery slightly to get the boat balanced. Check that the propeller cannot touch the water.
8. Now connect the electric motor to the battery as shown. The propeller should spin rapidly, making the boat move through the water. If the boat moves backward, then the propeller is spinning the wrong way. To make it turn in the other direction, change over the connections to the battery.
9. If the boat does not keep a straight course, bend the rudder slightly. If the boat turns to the left, bend the rudder to the right, and vice versa.
10. Experiment with different shapes and sizes of propeller and see which makes the boat go the fastest. You will probably get the best results if you use a **plastic propeller** designed for a model aircraft. Do not test the boat in a pond unless the water is very calm. Waves could make the boat tip over and dump the battery and motor into the water.

Take care!

When cutting and filing pieces of the soft-drink can, wear thick gloves to protect your hands.

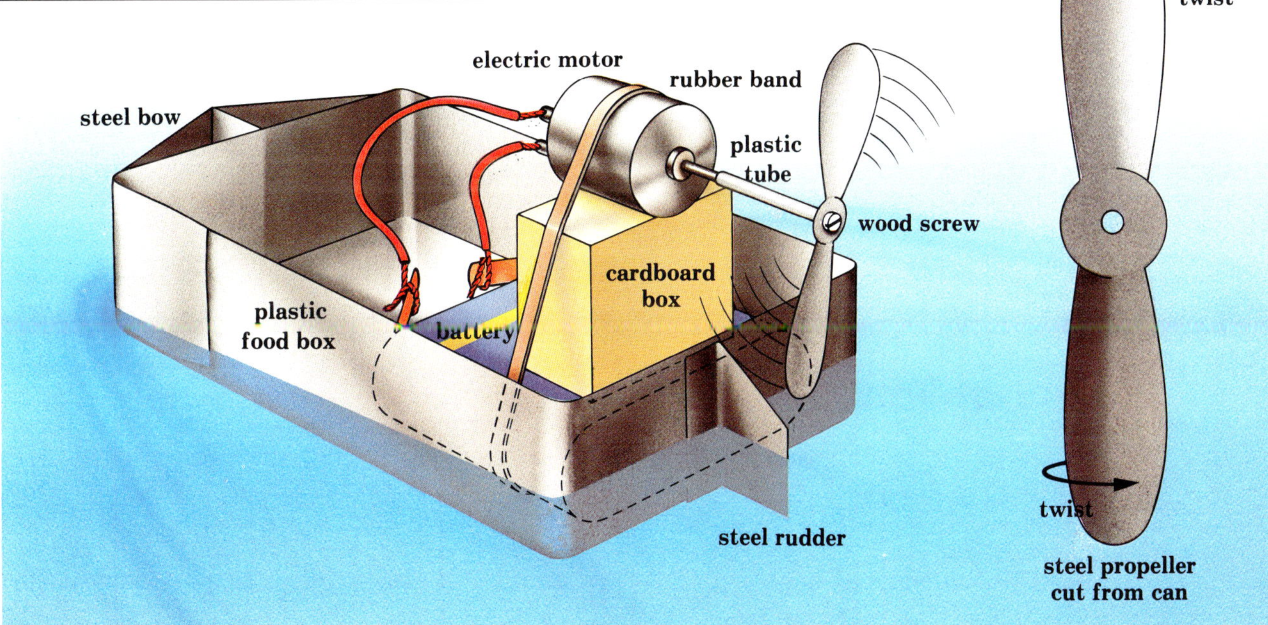

Crystal radio

You will need–

enameled copper wire
cardboard or plastic tube
tape
1 300 picofarad fixed
  capacitor
1 germanium or silicon
  diode
1 crystal earphone
1 wooden board
thumb tacks
1 staple
1 long metal rod

Early this century, crystal sets were used for listening to radio broadcasts. The crystal set is an extremely simple form of radio receiver. It requires no electricity supply of any kind.

How it works

A long piece of insulated wire is used as an antenna to pick up radio waves. The waves cause varying electric currents to flow down the wire. The currents pass through the crystal set and into the ground.

Inside the crystal set is a tuned circuit. This consists of a capacitor (condenser) and a coil. The tuned circuit is used to select the required signals from the mixture picked up by the antenna.

The electrical signals selected by the tuned circuit cannot be converted directly into sounds. This is because they alternate (change direction) rapidly— hundreds of thousands, or even millions of times every second. An earphone cannot reproduce vibrations at this frequency (rate).

Sound signals are obtained by connecting a diode between the tuned circuit and the earphone. The diode is an electronic component that passes current in one direction only. The earphone is able to respond to this one-way (direct) current. The result is that a thin plate in the earphone moves according to the strength of the received radio signals.

Procedure

1. For the tuning coil, you will need **enameled copper wire**. Its thickness is not important, but 22 gauge is recommended. The coil is wound on a cardboard or plastic **tube** measuring 1 to 2 inches (2.5 to 5 centimeters) across and about 4 inches (10 centimeters) long. To make the coil, start by attaching the end of the wire to one end of the tube with **tape**. Or make two small holes in the tube and thread the wire through them. Then wind the wire neatly around the tube so that the turns lie side by side. At intervals, twist the wire to form a "tail" protruding from the coil, and then continue winding. Make about six of these tails, which should be closer together near one end of the coil. The more turns on the coil, the greater the frequency range of signals that you will be able to receive. So make a coil with at least 100 turns.

2. Buy a **300 picofarad fixed capacitor** and a **germanium or silicon diode** suitable for use as a detector. To listen to the received signals, you will need a **crystal earphone**. (The more common low-impedance type of earphone is unsuitable.)

3. Fasten the parts to a **wooden board** and connect as shown. You will need **thumb tacks** and a **staple**. Scrape the enamel from the ends and tails of the coil before fixing them in place.

4. Run a wire (red in the illustration) out of the house

—if you are indoors—and attach it to a convenient high point. The branches of a tree would be suitable. This wire will act as the antenna. Another wire (here green) should also lead outdoors, where it is attached to a long **metal rod**. The rod must be hammered about 2 feet (0.5 meter) into the earth, where it will make the ground connection.

5. To search for signals, touch the free wire (here yellow) onto the coil connections, one after another. In this way, you should be able to tune in to some of the more powerful stations in your area. If the stations overlap too much, try connecting the antenna separately to a lower point on the coil.

Take care!

When attaching the wire to a high point as instructed in step 4, use a firmly positioned ladder, or ask an adult to do it for you.

Safety first!

Safety

The chemicals used in the projects can all be handled safely. Most are common household substances such as salt, baking powder, vinegar, etc.

When a project calls for an electricity supply, there is no danger of electric shock because a low-voltage battery is used.

A few projects involve the use of a flame or heat from a stove. In such cases, younger experimenters should be supervised by an adult, but the procedure is as safe as cooking.

Supervision

The projects have been graded according to the need for adult supervision. Where a project is marked with an ⊕, it means that, for complete safety, an adult should assist the young experimenter with some aspects of the project. In many cases, the adult's assistance will be limited to helping with some parts of the project, such as using a hammer and nails, and then letting the experimenter continue the project with only background supervision. Similarly, it may be necessary for an adult to handle matches or candles.

In other cases, the adult can prepare the materials which are needed for the experiment. For example, if the project includes accurate cutting out with a sharp knife, the experimenter need not handle the knife if the adult does the cutting out beforehand.

It is a good idea for an adult to be present when the project involves breakable objects like glass jars. A little guidance will minimize the risk of breakages.

Sharp edges

Wherever it is possible, the materials chosen for the projects are the safest ones that can be used. Sometimes, there is a choice of materials. For instance, the risk of sharp edges is reduced if you use plastic glass, but if you do use conventional glass, the chance of getting cut will be minimized if you are careful. First, protect your hands with gloves and use a small file or some sandpaper to remove any sharp edges from the glass. Handle the glass carefully and never put too much strain on it.

Similar precautions should be taken if the project involves cutting metal. Again, wear gloves and file off any sharp points. For extra safety, cover the cut edges with thick adhesive tape. If the project uses an empty tin can, try to find a can with a push-fit lid. This means you do not need to use a can opener, which will leave a sharp edge.

Chemicals

The chemicals used in the projects are all harmless, but they should still be treated with care. Keep each chemical in a labeled jar, and make sure that the jars are stored out of reach of inquisitive small children. Do not experiment with chemicals anywhere near food. Cover the worktable with old newspaper; it will catch any spilt chemicals and can be thrown away later. After the experiment, wash thoroughly the jars or dishes that have been used and pour the old chemicals down the sink, flushing them away with lots of water. Finally, wash your hands to remove any traces of chemicals on the skin.

Fire

You should take extra care with those projects which involve matches or candles. Organize the worktable so that there are no scraps of paper lying around and make sure that you are not wearing loose clothing, such as a tie or scarf which might accidentally catch fire. Check that you have all the materials needed for the experiment before you begin and arrange them on the worktable so that you do not need to reach over the candle. Always keep a pail of water close at hand just in case there is an accident.

Procedure

Before starting work on a project, it is important to read the instructions through to the end and to form a clear idea of what has to be done, and in what order. Materials and tools needed are listed in the margin and are spelled out in **bold type** when they are first mentioned. Make sure that everything is at hand when it is needed. Many of the projects or experiments can be carried out more smoothly if a little preparatory work, like weighing or cutting out, is done beforehand.

Science and the future

There are more and more opportunities for scientists in the modern world. Every year, new scientific discoveries help to change the world we live in. Most aspects of our life, including transport, entertainment, medicine and industry are changing rapidly because of the new inventions and discoveries that scientists are making. When you work on the projects in this book, you will learn many of the basic scientific principles which have helped important scientists to make their contribution to our world. Maybe one day, if you decide to become a scientist, you will join the great men and women of science who will create the world of the future. Who knows what you could achieve?

Words you need to know

In this book, you may find some words that you haven't seen before. These four pages explain as simply as possible what these words mean and will help you to understand exactly how to do the projects or experiments. Some words are special descriptions invented by scientists, and so are often very complicated to explain—in fact, whole books have been written about them! Of course, there isn't space in this book to include these very long explanations, but if you want to read more about any of them, ask your teacher or librarian to help you find a book.

Adhesive
Adhering to, sticky (like honey, for example)

Adjacent
Next to, lying beside

Air resistance
Resistance offered by air to a moving object

Antenna
Wire or rod for TV or radio reception

Attract
In physics, the tendency of objects to be drawn or pulled toward each other and to unite

Axis
The straight line, real or imaginary, about which an object revolves

Brittle
Easily broken or snapped, fragile

Buoying
Holding up, lifting

Capacitor
Condensor, conductors separated by insulator

Charge
In electricity, quantity carried or stored

Coil
In electricity, coiled wire for passage of electricity

Conduct
To transmit heat or electric current

Contract
To grow smaller

Conventional
Usual, customary, normal

Convert
To change or transform

Cycle
Recurrent period of events as a measure of time; in electricity, complete alternation of a current from positive to negative

Cylindrical
Shaped like a cylinder, long and round

Diameter
Straight line passing through center of circle or sphere

Dimension
Measurement

Disk
Flat, circular object

Dynamic
Relating to energy or power in motion; energetic, forceful or vigorous

Electromagnetism
Magnetism produced by an electric current

Ellipse
Regular oval made by a point moving in a plane

Embossed
Carved or molded in relief

Essential
Vitally necessary

Exhaust gas
Gases thrown out by combustion or fire

Expand
To grow larger

Felt
Cloth made from rolled wool or similar materials

Fins (aerial)
Sharp projecting surface to guide or stabilize rockets

Flexible
Easily bent

Frequency
The number of cycles per second of current

Frieze
Decorated band or strip around a wall

Galvanize
Stimulate with electricity; coat with zinc

Gauge
Standard or exact measure

Generate
To produce electricity or heat

Germanium dioxide
Grayish-white metallic element

Germinate
To sprout, shoot or bud

Globe
Perfectly spherical body

Gravity
Force with which the Earth attracts other objects toward it

Horizontal
Flat or level

Identical
Exactly the same

Immerse
To dip or plunge into a liquid

Incandescent
Glowing or shining with heat or light

Impact
Striking together, shock, strong influence

Insulated
Separated by non-conducting materials

Laminate
To form or press into a thin sheet or layer

Magnetic field
Space around a magnet where its influence can be felt

Miniature
On a very small scale, minute

Parallel
Straight lines that are a constant distance apart

Pitch
In music, that sound determined by the frequency of sound waves reaching the ear

Pivot
Shaft or pin on which things turn

Plumb line
Cord and weight used to determine vertical direction

Precise
Accurate, exact

Propulsion
Act of propelling or pushing forward

Ray
Beam of light or energy

Reaction
Opposite or responsive action to a force

Rectangle
Four-sided figure with four right angles

Register
To indicate on a scale

Resistor
Device having resistance to electric current

Rigid
Stiff, firmly fixed

Rotate
To move around an axis

Safety valve
Valve that opens to relieve excess pressure

Silicon
Nonmetallic element common in nature

Simulate
Imitate conditions of

Spiral
Winding around a fixed point or center and constantly receding from it

Spool
Reel for winding yarn or thread, or cylinder

Transparent
Allowing light through so objects can be seen distinctly

Trial and error
Learning from repeated experiments and failures

Vacuum
A totally empty space without air or gas inside

Vertical
Upright

Vein
The blood vessel that carries blood to the heart; thin tissue forming the framework of a leaf

Volt
Basic unit of electromotive force

Index